I See Cu

by Katherine Scraper

I need to know these words.

cube

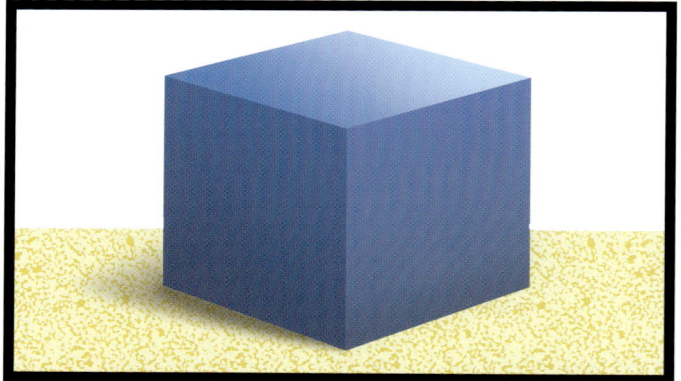

green

red

yellow

3

Look at this cube.
This cube is red.

Look at this cube.
This cube is blue.

Look at this cube.
This cube is green.

Look at this cube.
This cube is yellow.

Look at this cube.
This cube is white.

13

Look at this cube.
This cube is brown.

Is this a cube?

How are these cubes different?

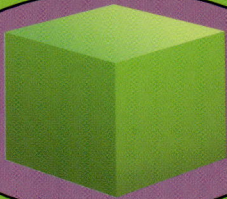

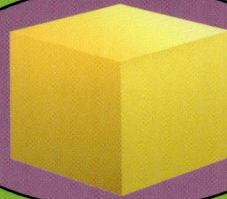

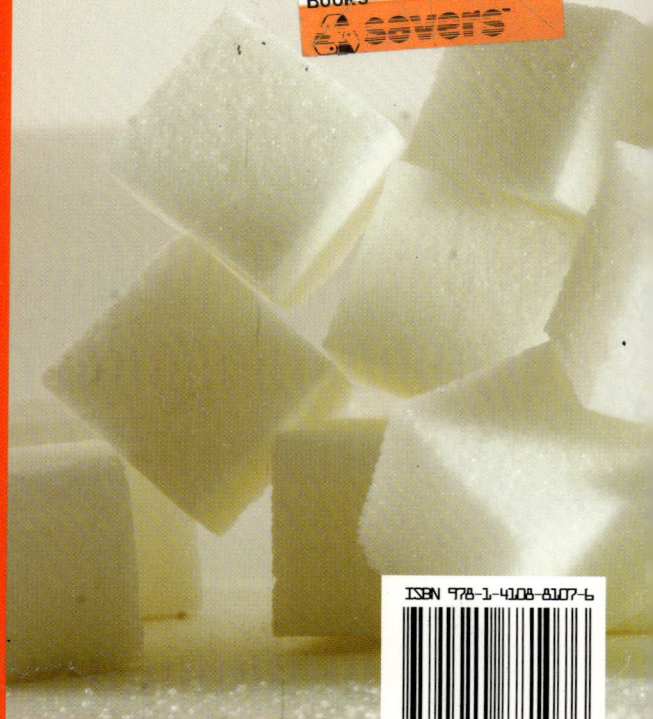

early Explorers™

Math

I See Cubes

Level B/2

Do you know what a cube looks like? All cubes are alike in some ways. All cubes can be different in other ways.

Theme:
Shapes

Other Early Explorers titles in this theme:

Benchmark
EDUCATION

BENCHMARK EDUCATION COMPANY

ISBN 978-1-4108-8107-6

9 781410 881076